MAP TECHNICAL PUBLICATION

Published by
MODEL & ALLIED PUBLICATIONS LIMITED
13/35 BRIDGE STREET
HEMEL HEMPSTEAD, HERTS.
1970

A Gallery of OLD TIMERS

51 pictures of Steam Wagons, Traction Engines, Fairground Rides, Ploughing Engines, Rollers and other old-timers.

by

ANTHONY BEAUMONT

Contents

31. **Foster SRL "Success" DCC No 14632. 1934.**
32. **Burrell S. tractor "Island Prince" DCC No 3868. 1921.**
33. **Fowler S. tractor DCC No 14412. 1918.**
34. **Fowler SC TE No 12761. 1911.**
35. **Ransomes Sims & Jefferies SC TE No 25542. 1912.**
36. **Ransomes Sims & Jefferies SC crane TE No 31066. 1921.**
37. **Burrell DCC TE No 4037. 1926.**
38. **Burrell SCC TE No 3918. 1923.**
39. **Marshall SC TE No 29944. 1900.**
40. **Clayton & Shuttleworth SC TE No 44104. 1911.**
41. **Allchin SC TE No 2146. 1922.**
42. **Aveling & Porter tractor (roller) No 9267. 1918.**
43. **Marshall SC roller Class Q. No 76797. 1923.**
44. **Aveling & Porter SC roller No 8361. 1914.**
45. **Aveling & Porter SC roller No 9155. 1921.**
46. **Wallis & Steevens 'Advance' roller No 8047. 1930.**
47. **Burrell SC roller No 4083. 1928 (last made).**
48. **Ruston & Proctor SC portable engine No 29011. 1905.**
49. **Steam cart by R. Soame. 1897.**
50. **Merryweather 'Gem' fire-engine No 2691. 1908.**

ABBREVIATIONS

TE	–	Traction engine
RL	–	Road locomotive
SRL	–	Showman's road locomotive
S tractor	–	Showman's tractor
SC	–	Single cylinder
DCC	–	Double crank compound
SCC	–	Single crank compound
NHP	–	Nominal horsepower (A maker's classification; the available brake horsepower is 5–6 times the NHP; even more for brief periods of work)
C	–	Approximately

Introduction

The enthusiasm for traction engines, allied steam vehicles and fairground amusements is not easily satisfied in one picture booklet. Those people who cannot tear themselves away from a working showmans engine, belted up to the dynamo, may well pass a steam wagon with hardly a second glance! Again, the student of showland history who is able to recite past owners and vicissitudes of a set of gallopers may not extend such devotion to the humble steam roller. Certainly, when fair organs play at rallies, they attract constant and close attention from ardent followers bent on tape-recording a favourite instrument or tune. Flourishing organisations exist which cater for this wide variety of interests. I name but one— The National Traction Engine Club, whose membership secretary is D. H. Ehlen Esq., 86 Kestrel Road, Kempshott, Basingstoke, Hants.

In this picture selection, I have aimed at a reasonable satisfaction of this combination of tastes which reflect the wider scope of rallies nowadays. I hope that the photographs will make a pleasant change from a catalogue of rally engines. Apart from a few copied illustrations, all the photographs are unpublished hitherto.

All enthusiasts who possess old and probably faded or tattered engine prints, can preserve valuable historical material by having photocopies made. A number of these old pictures are very clearly detailed, having been taken on large glass plates. Carefully-stored new negatives can always produce prints which are superior to the originals and the two Savage wagons illustrated have been "saved" in this way.

Repetition of my previous introductory summaries of engine types and uses makes tedious reading, however one "rings the changes". I think that the basic facts have now been well and truly covered in a wide variety of books during the past ten years; moreover, all my titles remain in print.

It is sufficient to record here that the traction engine displays have now continued for twenty years with steadily increasing enthusiasm on the part of owners and the general public. Traction engines and similar vehicles in "rally condition" are now valuable machines and in cash terms, a very far cry from the numerous disused engines in good condition which could be bought for little more than scrap prices in the late 1940s. Indeed, the humble traction engine in these changed times appears to have joined "Old Masters", antique silver and latterly, veteran motor cars as an "inflation hedge". This year, a small steam tractor offered by auctioneers of world renown, fetched ten times its original new price. Certainly the days of the keen but impecunious steam enthusiast are now long gone. I am reminded of a friend's father who once *paid* gipsy dealers to remove a whole cartload of traction engine oil lamps.

Irrespective of engine finances, the augury for steam preservation is a happy one. The demise of the steam railway locomotive has concentrated interest in the remaining steam vehicles to an extent never envisaged by the wildest optimist in the 1940s. The production of "The Traction Engine Register" in 1968, listing some 2000 existing steam vehicles — all but a very few situated in the U.K. — is an amazing fact. I illustrate the position by a quotation from "Traction Engines Not So Well-Known" by R.H. Clark Esq., in "The Model Engineer" for 1949. "At the present rate of destruction it will only be a very few years before most traction engines are gone, and examples of the larger makers will be the last to go, but before then the smaller number of those not so well-known makers will have gone to the scrap-heap".

After years of Herculean efforts by purchasers of derelict and badly neglected engines, it would seem that the sands of time have almost, if not completely, run out for "discovering" restorable machines. A recent example of a courageous but abortive attempt to save an engine resulted in a Burrell showmans being finally cut up after a few parts were saved for possible spares. This engine had been lying untouched for forty years exposed to salt air except for the covered dynamo. We can believe that there was insufficient metal left to make an engine.

I have mentioned the subject of boilers elsewhere. While it is inappropriate to "lay down the law" from this platform, it has seemed essential to experts that the question of regular boiler testing should be uppermost in owners' minds. I reiterate this weighty opinion as sound commonsense, applicable to all engines which are steamed, however infrequently. The consequences of a major boiler failure at a crowded rally can be imagined. The "factor of safety" in every *new* boiler is an insurance against the danger of some totally unforeseen defect developing, but this factor should never be relied upon to replace proper testing, care and maintenance. In the days of hard-worked traction engines, their boilers, particularly the thin-walled tubes, fared much better than in engines which are infrequently used and sometimes stand with undrained boilers.

Accounts of fair organs have appeared in other books and the increasing popularity of the older fairground rides is most pleasing to all of us having memories of fairs up to the mid-1920s. Regrettably, none of the largest scenic railways exists, but a number of "Old Time Steam Fairs" are being staged, featuring a surprising variety of authentic steam-driven rides both in showland and private ownership. Very "well-bred" wood horses for gallopers are being carved here as replacements for a number of 3-abreast sets; a new set of gallopers for inside use has been made up recently and various replacement parts are still in demand and are supplied by the makers. The interest in organs is such that a skilful and musical gentleman has actually constructed a complete instrument for himself and I am able to illustrate it.

Since 1865 when S.G. Soame in the village of Marsham, Norfolk, made the first recorded steam engine to drive a fairground ride, immense mechanical and technical progress has taken place. In these days of electronic and supersonic wizardry, the traditional bond of steam and ponderous mechanical amusements endures as a tribute to our old-time engineering and artist-craftsmen. They would have been well-pleased to know that their smooth-riding gallopers would still be turning after eighty years of pleasure-giving.

Anthony Beaumont,
South Wootton,
King's Lynn.

May 1969.

ACKNOWLEDGMENTS

The following firms and individuals have helped in making the variety of pictures. I extend my thanks to all who have assisted.

Ransomes, Sims & Jefferies Ltd.
J. Fowler (Leeds) Ltd.
Wallis & Steevens Ltd.
Fodens Ltd.
Savages Ltd.
A. Bloom, Esq.
R. Butterfield, Esq., Kericho, Kenya.
R.H. Temple, Esq.
F. Howling, Esq.
G. Kelly, Esq., Co. Wexford.
H. Pye, Esq.
J. Desborough, Esq.

All the photographs are published for the first time in book form and 35 are unpublished hitherto.

1

A small Savage set of 3-abreast horses and giant poultry, made 1895. Tidman centre engine not geared up in 1968. Bressingham 1968.

2

The Bruder (restored by V. Chiappa) 46 Key organ and details of the Bressingham set.

3

Savage 3-abreast set with an 1887 Savage centre engine using V-belt drive instead of spur gearing. Thursford 1968.

4

Chairoplanes at Thursford 1968. Introduced in the early 1920s, Chairoplanes were usually fitted with a small organ.

5

Details of H. Lee's Savage Steam Yachts 1900, showing the divided crankshaft of the duplex cylindered engine and chain and rod yacht swinging gear. Stamford 1964.

6

A Juvenile hand-turned ride by Orton & Spooner 1895. The carved work is the original. Stamford 1967.

7

An excellent reproduction of an old time fair organ – a 46 key instrument made by A. Cragg in 1967 – at Kegworth in 1968.

8

Bruder 46 key organ imported 1938, restored by V. Chiappa 1968.
Kegworth 1968.

9

Chiappa 98 key organ, rebuilt in 1958 from 98 key Marenghi, new in 1908 as 104 key for Pat Collins bioscope. Used from 1911 in scenic railways as 98 key.

10

Bartlett's 3-abreast swinging (not crank overhead) horses and chariots shown C. 1894. The SRL extreme right is almost certainly Burrell's "Pride of the South" No. 1470, 1890 with dynamo added.

Savage "Sea-on-Land" from 1905 catalogue The boats were pitched by eccentrics and links on geared-down counter-shafts driven from the track wheels.

12

Aveling & Porter SC TE with twin fly-wheels shown with REs of balloon section, Brompton barracks, Gillingham C. 1885. Driven by W. Pye, Lidsing. Note the cyclist.

A Savage OA wagon at a King's Lynn millers in 1910, showing steam collecting drum of the water tube boiler.

14

Nearside view of similar Savage wagon after picking up water in 1907. The reverse lever connection to the hollow crankshaft is shown.

15

Works photograph of Savage wagon with platform body and steel wheels. Nominal load was 6 tons. The enclosed compound cylindered engine is well shown.

Wallis & Steevens over-type wagon, DCC, with wood wheels, from a works' photograph C. 1910.

17

Wallis & Steevens 15-ton roller when new at the Basingstoke works C. 1910. Note the thickness of the rolls

Wallis & Steevens roller showing its scarifying capabilities in an Edwardian setting C. 1907.

Foden DCC RL of 8 NHP in the early 1900s; hauling power 45 tons nominal.

Fowler side drum ploughing engine DCC, Class B.4 made in 1900. The cable guide pulleys are beneath the tender.

21

Fowler SC ploughing engine, Class K.5 with automatic cable coiling gear. First made in 1895.

Improved Single-Cylinder Self-Moving Ploughing Engine

Standard Type

Road-Gear Side.

Foster light agricultural TE of 5 NHP with two speeds; a makers illustration C. 1910.

23

Foden 'Speed 6' undertype wagon with duplex cylinders mounted transversely and two speed shaft drive. The final type in 1931.

24

Ransomes Sims & Jefferies No 2 overtype articulated wagon; a works' photograph in 1922 of a unique design of this firm.

25

Marshall
Colonial tractor,
SC single speed
with separate
steam dome.
Shown in Kenya
1967.

26

The sole survivor of Brown & May's showmans engines – "General Buller" No 8742 DCC, 7 NHP 1916. Used for threshing until early 1950s.

27

McLaren SC TE No 163 of 1883 or 1884; three shaft two speed, with an owner-modification of a lengthened boiler (between cylinder and chimney). Eire 1967.

A similar McLaren No 127 made 1882, seen threshing at Paverham in 1967.

29

Marshall SC TE No 69254 1921, threshing at Thursford 1967. The engine's governor deals with variable loads, allowing the driver to leave the footplate.

Ruston & Proctor SC TE No 113043 1920, driving a baler, fed with straw from an elevator in the set-up shown in the previous picture.

31

Foster SRL "Success" No 14632 DCC, '65 BHP' class, three speed. Made in 1934 for Amusement Caterers Hibble & Mellor, Nottingham. Shown here at Kilverstone, Thetford 1958.

Burrell S tractor "Island Prince" No 3868 DCC, 4 NHP three speed, built 1921 as a normal tractor and formerly with Arnold's Amusements in the Isle of Wight.

33

Fowler S tractor "Tiger" type 3½ NHP No 14412, three speed, DCC, made in 1918 and seen here at Stamford 1967.

Fowler SC TE No 12761 7 NHP, made in 1911; once owned in York-shire and pict-ured here at Market Bosworth in 1966.

Ransomes Sims & Jefferies SC TE No 25542, built 1912 and new to Wisbech Cambs. Used for threshing up to 1958. Market Bosworth 1966.

Ransomes crane DCC TE 6 NHP No 31066 1921 with steam steering gear operated from the crane winch shaft. A Fowler 'Leeds Arms' decorates the smoke-box door.

37

Burrell DCC TE No 4037 5 NHP made 1926 and the final form of the firm's design. Shown here at Raynham Park 1963.

Burrell TE single crank compound No 3918, 5 NHP made in 1923. Only seven more SCC machines were made after this engine. Revesby Park 1963.

39

Marshall SC TE No 29944, 6 NHP built in 1900, with two speeds and rear wheel brake. Photographed at Carrington Park 1964.

Clayton & Shuttleworth SC TE No 44104, 7 NHP weight 10 tons made in 1911. Now named 'Enterprise' and shown here at Market Bosworth 1966.

41

Allchin SC TE No 2146, 6 NHP two speed, built in 1922 and photographed in rain at Stamford 1967.

42

Aveling & Porter tractor DCC No 9267, 4 NHP made 1918 and converted from a roller 1953, hence the non-standard wheels. The author is included here at Market Bosworth in 1966.

Marshall roller SC 10-ton Class S No 79597, 6 NHP two speed made in 1925. Fitted with Marshall's radial valve gear using only one eccentric on the crankshaft. Revesby 1963.

Aveling & Porter SC roller 8-ton No 8361, 6 NHP. Made 1914 and formerly owned by Surrey CC at Richmond. The rear roll hub plate is a maker's repair. Terrington 1969.

45

Aveling & Porter SC roller 8-ton No 9155, 5 NHP built 1921. New to Pulham, Norfolk and shown here at Thursford 1967.

46

Wallis & Steevens 'Advance' duplex cylindered piston valve roller 6-ton No 8047, 5 NHP 1930. Used for airfield construction 1939-45 and shown here at Market Bosworth 1966.

47

Burrell SC
roller No 4083,
8-ton 5 NHP,
two speed; made
1928 and new
to East Suffolk.
The last Burrell
roller made.
Woodton 1960.

48

Ruston & Proctor portable engine SC No 29011, 8 NHP made 1905 with patent steam-jacketed stay from cylinder to crankshaft bearing. Seen at Kegworth 1968.

Unique steam cart made by R. Soame of Marsham, Norfolk in 1897. Restored in late 1950s. Vertical solid fuel boiler steams SC engine with final belt drive. Woodton 1960.

Merryweather 'Gem' fire engine No 2691 made 1908 with duplex cylinders directly coupled to pumps via slide cranks. Stamford 1964.